Andy Lenz

Tornadogenesis. What causes a Tornado?

GRIN Publishing

Bibliographic information published by the German National Library:

The German National Library lists this publication in the National Bibliography; detailed bibliographic data are available on the Internet at http://dnb.dnb.de .

Imprint:

Copyright © 2015 GRIN Verlag GmbH
Print and binding: Books on Demand GmbH, Norderstedt Germany
ISBN: 978-3-656-94355-6

This book at GRIN:

http://www.grin.com/en/e-book/296265/tornadogenesis-what-causes-a-tornado

GRIN - Your knowledge has value

Since its foundation in 1998, GRIN has specialized in publishing academic texts by students, college teachers and other academics as e-book and printed book. The website www.grin.com is an ideal platform for presenting term papers, final papers, scientific essays, dissertations and specialist books.

Visit us on the internet:

http://www.grin.com/

http://www.facebook.com/grincom

http://www.twitter.com/grin_com

Tornadogenesis: What Causes It

Andy M. Lenz

Lausanne Collegiate School

Tornadoes are a ruthless, frightening force of nature. They are capable of wiping entire towns off the map and killing scores of people in their way. Their sheer power is extremely difficult to measure even with the most cutting edge weather instruments because the instruments themselves often cannot survive the impact of a tornado. People have feared and been captivated by this fascinating phenomenon for eternities. Because of their destructive power, intense research has been conducted in recent years to understand tornadoes. Millions of dollars have been spent funding studies and research. The government itself even funding a massive field research project on tornadoes called Vortex 2, which was the largest project ever to study tornadoes. All of this research has led to significant breakthroughs in how tornadoes form and the conditions that are favorable for their formation. Once a seemingly indecipherable force, tornadoes now have known factors that lead to their formation. These factors include, temperature, atmospheric instability, moisture, atmospheric forcing, and wind shear, among various other factors. One factor stands out from the rest, and that is wind shear. The presence of wind shear shows tremendous promise in identifying conditions favorable for the development of tornadoes. In fact, it seems to be the most pivotal condition needed for tornadoes to form due to multiple compelling reasons. Wind shear is the most important factor in tornadogenesis because it causes storms to rotate, sustains storms, and is present in almost all tornadic events.

Wind shear is absolutely critical in the development of rotation in storms and for that reason, it is also critical for the development of tornados themselves. Wind shear is the differentiation between the wind speed and or direction at different altitudes directly at and above a constant point on the surface of the Earth. There are two basic types of wind shear: vertical speed shear and vertical directional shear. Vertical speed shear is caused by the differentiation between the speeds of the wind at different altitudes. Vertical directional shear is caused by the differentiation between the directions at which the wind is blowing from at different altitudes. Both classifications of wind shear can be present simultaneously in the atmosphere at a given location or be independent of one another as well.

Wind shear initiates rotation in thunderstorms through the occurrence of a well-documented process. "It is widely accepted that vertical vorticity initially arises within thunderstorm updrafts as a result of tilting and subsequent stretching of horizontal vorticity associated with mean vertical wind shear." (Markowski & Richardson, 2008). The author clearly attests the practically indisputable nature role of wind shear in the development of rotation in thunderstorms. Coming from an expert, this statement speaks volumes, as it proves the thinking of the scientific community is in line with this specific aspect of this paper's argument.

As to how wind shear compares to other factors of tornadogenesis, the evidence is striking and abundant. "Two parameters seem to offer the most promise in discriminating between non-tornadic and tornadic supercells: (1) boundary layer water vapor concentration and (2) low-level vertical wind shear (Markowski & Richardson, 2008). This is another strong example of how important wind shear to the development of tornadoes. Not only does wind shear cause tornadic rotation, but it also can be used to distinguish tornado producing thunderstorms from non-tornadic thunderstorms, a testament to its importance. Another reason that wind shear is key to the development of rotation in tornadic thunderstorms has to do with helicity. Helicity is a calculation the measure the ability/potential of a thunderstorm to rotate. It is determined by wind shear, as one might imagine. "Very large helicity in the lowest 1km is particularly favorable for tornadoes (Davies-Jones, 2006). Once again, wind shear is proved to be an impetus for tornadic rotation. Because of all the compelling evidence, there is no question that wind shear initiates the rotation in tornadic thunderstorms and due to that, it is a pivotal factor in tornadogenesis.

Wind shear does more than just create rotation in thunderstorms, it sustains them as well. While the development of rotation is obviously an important part of tornadogenesis, a tornado needs a large, strong, and fast updraft that is sustainable for a prolonged period of time. If the updraft is too small or too weak it would be very difficult for a tornado to develop. In addition, that updraft not only needs to be powerful, but also long lasting because if the thunderstorm dissipates too quickly, a tornado will not even get the opportunity to form. Wind shear sustains thunderstorms because "Strong upper tropospheric winds evacuate mass from the top of the updraft. This reduces precipitation loading and allows the updraft to sustain itself (Haby). Undoubtedly, this information supports the claim about sustentation of thunderstorms by wind shear. Furthermore, "Modeling experiments indicate that low-level curvature shear and storm relative helicity interacting with deeper tropospheric shear can enhance strongly the intensity of an updraft (Johns, Davies, & Leftwich, 2013). There is no question about the vitality of wind shear to sustaining strengthening thunderstorms, allowing them to produce tornadoes. For proof of these concepts, the evidence couldn't get much more convincing. "Fawbush and Miller (1954) identified environmental conditions that are conductive to large, long-lived tornadoes. The sounding generally has large convective available potential energy (CAPE) and strong wind shear associated with winds veering and

3

increasing with height" (Davies-Jones, 2006). The research conducted by Fawbush and Miller leaves no room for doubt about the validity of the role of wind shear in sustaining tornadic thunderstorms. Overall, all the information provided definitively points to the importance of wind shear in maintaining tornadic thunderstorms and thereby the development to tornadoes themselves.

Wind shear may cause storms to rotate and sustain them as well, but the most definitive evidence of the role of wind shear in tornadogenesis comes from the data collected from tornadic events. This data paints a convincing picture of how important wind shear is to tornadogenesis. In order to accurately determine whether wind shear is the most imperative factor in the development of tornadoes one must not only prove how wind shear leads to their formation, but also that it is present in practically all tornadic events. Otherwise, it would be difficult to reach such a conclusion. As for the data, "The 0-6k km BWD [wind shear] was found to contribute equally to all three significant tornado regimes." (Garner, 2013). The significance of this research is undebatable. If wind shear contributes equally to all tornadic events then it is next to impossible to dispute its significance. While other factors may be inconsistent, showing up in different places at different times in different amounts, the role of wind shear remains steadfast, alluding to the claim that it is the most influential factor in tornadogenesis.

For the next section, I conducted original research and compiled the data to further illustrate the fact the wind shear is present in virtually all tornadic events. The way in which I gathered the data follows. First, I went the Storm Prediction Center's Severe Weather Events Archive. I decided to analyze the all days during the 2014 calendar year that the Storm Prediction Center issued a moderate or high risk of severe thunderstorms. However, to avoid analyzing severe weather events that carried over from one day to the next and make sure all the events were independent outbreaks, I omitted events that met the criteria if they occurred less than three days after another event. Consequently, in the case of more than one event in less than three days, I chose the most recent event and omitted the next. Because I decided to obtain my data this way, there were twelve events that met my criteria. I chose to gather the data in the way that I did because using only the most significant severe weather events (moderate or high risk) is more practical than combing through all the countless lesser severe weather events and because it focuses more on the conditions in an environment where tornadoes are more likely to form. In other words, significant severe weather are more likely to produce tornadoes than minor severe weather events and since the study focuses

4

environments favorable for tornadoes, using significant severe weather events is the best option. I chose to only include significant severe weather events from the 2014 calendar year also because it was more practical, but for a couple of other reasons as well. First off, the 2014 calendar year is the most recent full year of data. And second, by obtaining data from one calendar, I capture severe weather event from all throughout the four seasons, not just spring, which is the primary severe weather season. The analysis of my follows in the next paragraph. One will find that the data collected strongly supports my argument.

I analyzed the data from the twelve significant severe weather events that met my criteria for two for two parameters- effective bulk shear and effective storm relative helicity. The parameters are both obviously measures of wind shear. I thought these parameters would best encapsulate the environment of tornadic events from a wind shear perspective. Here is the raw data:

Event 1: October 13, 2014   -

50-60 kts

400-500 m2/s2

34 tornadoes

Event 2: Jul 27, 2014-

50-60 kts

200-300 m2/s2

11 tornadoes

Event 3: July 8, 2014-

50-60 kts

100-200 m2/s2

14 tornadoes

Event 4: Jun 30, 2014-

50-60 kts

400-500 m2/s2

26 tornadoes

Event 5: Jun 16, 2014-

50-60 kts

400-500 m2/s2

35 tornadoes

Event 6: Jun 3, 2014-

60-70 kts

500-600 m2/s2

16 tornadoes

Event 7: May 11, 2014-

70-80 kts

500-600 m2/s2

36 tornadoes

Event 8: May 08, 2014-

30-40 kts

100-200 m2/s2

9 tornadoes

Event 9: Apr 28, 2014-

60-70 kts

400-500

121 tornadoes

Event 10: Apr 13, 2014-

30-40 kts

200-300 m2/s2

8 tornadoes

Event 11: Apr 03, 2014-

50-60 kts

300-400 m2/s2

16 tornadoes

Event 12: Feb 20, 2014-

60-70 kts

400-500 m2/s2

34 tornadoes

The first thing to note from the data as that all the significant severe weather events analyzed had at least thirty to forty knots of effective bulk wind shear. This is imperative because it shows that no matter what the state of the atmospheric environment is during a tornadic event, there is always a sizable amount of effective bulk wind shear, or at least in the events that I studied there was. The maximum amount of effective bulk wind shear observed in the dataset during any particular event was seventy to eighty knots of wind shear. This is probably the case be there can be physically so much effective bulk wind shear can be present at a given time and that is likely close to the max. Another critical detraction from the dataset is that in general, the number of tornadoes reported on a given day increases as the amount of effective bulk wind shear increases and the number of tornadoes reported decreases as the effective bulk wind shear goes down. For example, the two events with the least reported tornadoes, events eight and ten, had the lowest amounts of effective bulk wind shear by far at 30 to 40 knots. The next lowest amount of effective bulk wind shear was 50 to 60 knots. Also, the four events with the most effective bulk wind shear, which account for approximately 33.3% of the data, had approximately 57.5% of the tornadoes. This is key because it proves how profound an impact effective bulk wind shear has on the development of tornadoes. Not only is effective bulk wind shear present during all the significant severe weather events, but the amount this wind shear dictates how many tornadoes form on a given day. This further cements the clear cut role wind shear plays in tornadogenesis.

As for the effective storm relative helicity, the data as striking and consistent with my claim as the data for the effective bulk wind shear. The first correlation I noticed in the data on the storm relative helicity when I analyzed has to do with the minimum amount present. All of the significant severe weather events that I sampled had at least 100-200 m2/s2 of effective storm relative helicity. This proves that there is always a meaningful amount of effective storm relative helicity present during significant severe weather events, or at least in the events I sampled. This is consistent with the data from the effective bulk vertical wind shear. Both measures of wind shear appear to always be present in the data I analyzed. This speaks volumes about and further strengthens my claim that wind shear is present during all tornadic events. The maximum amount of effective storm relative helicity I observed in the data that I analyzed was 500 to 600 m2/s2. Once again this is likely due to the fact that is only physically possible for there to be so much effective storm relative helicity at a given time in a given place. The last extremely noteworthy pattern that I noticed in the dataset was the correlation between the amount of effective storm relative helicity and the number of tornadoes reported on the given day. The seven events with the most effective storm relative helicity, which account for approximately 58.3% of the data, had approximately 83.8% of the tornadoes. This continues the definitive trend from all of my previous analysis of the data that clearly shows that wind shear, in its various forms, is strongly linked to the development of tornadoes. Just like with the effective bulk wind shear, this proves that the effective storm relative helicity directly influences the number of tornadoes that form on a given day. Without a doubt, wind shear is the most important factor in tornadogenesis.

Wind shear is clearly the definitive indicator in tornadogenesis. It leads the development of tornadoes in many ways, It causes storms to rotate, it sustains them and is present is almost all tornadic events. The evidence supporting this is well-documented and unequivocal. There is no shortage of information that proves this point. In addition, the role of wind shear is so accepted in the development of that there is virtually no evidence to discredit it. Not only do expert opinions support this claim, but original supports it as well. Wind shear simply is the most important factor in tornadogenesis. The evidence is too overwhelming to believe otherwise. Given the epic destructive power of these storms, it is quintessential that tornadoes are understood. Without knowledge of tornadogenesis, humanity would be left helpless at the mercy of these storms. The situation would be similar to earthquakes. One never knows when one will strike, but when they do, there is a disaster. Imagine over a thousand of these unpredictable earthquakes every year. That would be the reality for tornadoes without knowledge of their formation. Maybe someday, with continued research and studies, meteorologists might finally complete the puzzle and eliminate the unknowns.

References

Center, S. P. (n.d.). *SPC Severe Weather Events Archive.* Retrieved from Storm Prediction Center: http://www.spc.noaa.gov/exper/archive/events/

Davies-Jones, R. (2006, January 31). *TORNADOGENESIS IN SUPERCELL STORMS – WHAT WE KNOW AND WHAT WE DON'T KNOW.* Retrieved from National Oceanic and Atmospheric Administration: ftp://www.star.nesdis.noaa.gov/pub/smcd/spb/lzhou/AMS86/PREPRINTS/PDFS/104 563.pdf

Garner, J. M. (2013, August 6). *A Study of Synoptic-Scale Tornado Regimes.* Retrieved from Electronic Journal of Severe Storms Meteorology: http://www.ejssm.org/ojs/index.php/ejssm/article/view/119/90

Haby, J. (n.d.). *Wind Shear.* Retrieved from Weather Predition Education: http://www.theweatherprediction.com/severe/ingredients/windshear/

Johns, R. H., Davies, J. M., & Leftwich, P. W. (2013, March 18). *Some Wind and Instability Parameters Associated With Strong And Violent Tornadoes: 2. Variations in the Combinations of Wind And Instability Parameters.* Retrieved from Wiley Online Library: http://onlinelibrary.wiley.com/doi/10.1029/GM079p0583/summary

Markowski, P. M., & Richardson, Y. P. (2008, September 19). *Tornadogenesis: Our current understanding, forecasting considerations,.* Retrieved from Penn State University: http://www.meteo.psu.edu/~pmm116/pubs/2009/MR09ATMOSRES.pdf